Hydrogen Safety

Topical Conference at the 2022 AIChE Spring Meeting and
18th Global Congress on Process Safety

San Antonio, Texas, USA
10 – 14 April 2022

ISBN: 978-1-7138-6822-4

Printed from e-media with permission by:

Curran Associates, Inc.
57 Morehouse Lane
Red Hook, NY 12571

AIChE
120 Wall Street, FL 23
New York, NY 10005-4020

Phone: (800) 242-4363
Fax: (203) 775-5177

www.aiche.org

Additional copies of this publication are available from:

Curran Associates, Inc.
57 Morehouse Lane
Red Hook, NY 12571 USA
Phone: 845-758-0400
Fax: 845-758-2633
Email: curran@proceedings.com
Web: www.proceedings.com

TABLE OF CONTENTS

638956 1. Properties of H2 .2.H2 Safety 4. MOC Consideration for H2service.5. Hydrogen Embrittlement 6. High Temperature Hydrogen Attack.7. H2 Accidents 8. Lesson Learned.

Abstract:

HYDROGEN SAFETY:

1. Properties of hydrogen - introduction of hydrogen to the participants including but not limited to physical properties

2. Hydrogen Safety : element hydrogen is not easily catching fire as well its fire cannot be easily detected by naked eyes. Safety considerations will be presented.

3. MOC Consideration of H2 service- material when comes in contact with H2 can cause leak ofH2. the operating conditions and criteria will be presented.

5. Hydrogen embrittlement - industrial case studies will be discussed

6. High temperature hydrogen attack- pipe network portions impact due to hydrogen will be presented here.

7. H2 Accidents: Incidents during handling and using hydrogen will be shared.

8. Lesson Learned: learning during use of hydrogen will be elaborated to upcoming hydrogen projects.

639515 Hydrogen Safety and Quantitative Risk Assessment – Key Knowledge Sharing

Abstract:

Hydrogen has significantly different physical and chemical properties to conventional natural gas and other hydrocarbons. Hydrogen storage, distribution and applications pose unique challenges due to hydrogen's wide flammable range, low minimum ignition energy, high buoyancy, wide range of explosion consequences, and its ability to embrittle metals. In addition, liquid hydrogen represents cryogenic hazards and challenges due to the low temperature it requires.

Hydrogen has a much lower density than natural gas so a H2 leak may disperse more rapidly than escaped natural gas. When released, the buoyancy of hydrogen enhances mixing with air and dilution, and this typically implies shorter dispersion distances (i.e., distance to flammable limits tends to be less in comparison to hydrocarbons).

Hydrogen explosions range from minor consequences to extreme outcomes for relatively small changes in ventilation, outflow rate, confinement, and congestion. Hydrogen detonations are entirely credible, even for small release scenarios, and are one of the worst-case events resulting from a H2 release. A detonation results in a shock wave that has great potential for causing personnel injury and/or property damage.

This paper provides an overview of hydrogen hazards and properties, focusing on dispersion and flammability (fires and explosions). In addition to a summary of a methodology to conduct H2 Quantitative Risk Assessment (QRA). H2 QRA methodology is similar to the one applied to hydrocarbons; however, some adjustments need to be performed, especially for estimating the leak frequency of release events. Nowadays, there is significant uncertainty to estimate the frequency of severe H2 explosions.

For hydrogen to gain broad acceptance and adoption, industry and regulators will need to establish robust safety standards and methods to quantify the risk associated with production, transportation, and usage cases, just as done for other potentially hazardous substances.

639351 A Foundational Framework for Reliability Data Collection for Hydrogen Systems

Abstract:

The international community has embraced the concept of mitigating climate change by transitioning to an energy system with reduced greenhouse gases and particulate emissions, as well as more sustainable energy production methods. Hydrogen has become a candidate for this transition by offering ways to decarbonize various sectors such as long-haul transport, chemicals production, and iron and steel. This is mainly due to its high energy content and clean-burning properties. It is widely recognized that safety and reliability are key issues for the development, commercialization and implementation of technologies or infrastructure. Thus, the wider adoption of hydrogen in various economic sectors, such as fueling stations, requires that the risks associated with it be rigorously investigated and quantified. Quantitative risk assessment (QRA) is an important tool in the development of risk-informed safety codes and standards that aim to reduce risk in engineering systems, including hydrogen fueling stations. QRA activities are instrumental to the development of protective and preventive measures that would make the commercialization and safe use of hydrogen fueling stations more feasible. Enhancing the safety of the systems would aid in gaining public trust and encourage policymakers to adopt regulatory frameworks, easing the way for their deployment on a wider scale.

Rigorous QRA needs reliability data, but current hydrogen safety databases are not in a conducive format for use in QRA. A literature review of articles focusing on hydrogen fueling stations found that lack of reliability data was the most common knowledge gap in the field. In this work, current hydrogen safety data collection tools are examined to assess the state of hydrogen-specific reliability data. These tools include H2Tools, Hydrogen Incident and Accident Database (HIAD), the National Renewable Energy Laboratory's Composite Data Products (CDPs), and the Center for Hydrogen Safety (CHS) H2 Equipment & Component Failure Rate Data Submission Form. Next, best practices are examined to determine the state-of-the-art in reliability data collection in analogous industries, specifically oil and gas in the form of the Offshore and Onshore Reliability Data (OREDA) project. The analyses of collection tools and best practices are combined to determine missing types of data and to create a reliability data collection framework for hydrogen fueling stations. This framework aims to present a standardized component hierarchy, failure mode taxonomy, and the necessary high-level elements for robust reliability data collection. This work lays the foundational basis for collaborative hydrogen reliability database that is conducive to QRA use, thereby working towards the goal of aiding wider adoption of safe hydrogen technologies in the economy.

640170 Ignition of Large-Scale Vapour and Two-Phase Hydrogen Jets: Field Data and Modelling

Abstract:

There is continued focus on the role of hydrogen in the energy transition with associated safety questions. Since hydrogen is highly flammable, fires are one of the main associated hazards. Depending on the application, hydrogen could be present as vapour or as cryogenic two-phase/liquid. However, field-scale data for vapour is limited and scarce for liquid/two-phase releases, making model development and assessment challenging. The current paper therefore examines field tests involving jet fires from large-scale releases of vapour and two-phase hydrogen. Analysis of field data is presented together with predictions from both established and recently developed consequence fire models.

Hydrogen jet fires are highly complex phenomena and non-linear in scale. This means that it is difficult to predict how large jet fires may behave based on small-scale lab experiments, and the current work therefore consider larger scale ignited releases. Two such sets of experiments were carried out by DNV at the Spadeadam test facility in the UK: ignited releases of hydrogen vapour (2008) and two-phase (2019). For the vapour jet fires we show how the traditional cone jet fire model of Johnson significantly underpredicts the radiation hazards, while the recent Miller model provides much more accurate radiation predictions for these large jet fires. For the more recent ignited two-phase hydrogen releases, we provide an analysis of the measured radiation data that should benefit analysts looking to validate fire models. Recommendations are made for choice of model for two-phase hydrogen jet fires, and particular considerations are made with regards to the complex downward impinging release experiment.

Overall, this contribution aims to make certain field-scale hydrogen jet fire data more accessible, and to provide updated guidance for the modelling of hydrogen jet fires.

NOTES

NOTES

NOTES

NOTES

NOTES

NOTES

AUTHOR INDEX